SCIENTIFIC AMERICAN EDUCATIONAL PUBLISHING

SCIENTIFIC AMERICAN INVESTIGATES SPIDERS

FUNNEL-WEB SPIDERS

BY ROXANNE ALDOLPHO

Published in 2025 by The Rosen Publishing Group
in association with Scientific American Educational Publishing
2544 Clinton Street, Buffalo NY 14224

Library of Congress Cataloging-in-Publication Data
Names: Aldolpho, Roxanne, author.
Title: Funnel-web spiders / Roxanne Aldolpho.
Description: [Buffalo, NY] : Scientific American Educational Publishing, an imprint of Rosen Publishing, [2025] | Series: Scientific American investigates spiders | Includes bibliographical references and index.
Identifiers: LCCN 2023059123 | ISBN 9781725351240 (library binding) | ISBN 9781725351233 (paperback) | ISBN 9781725351257 (ebook)
Subjects: LCSH: Agelenidae–Juvenile literature. | Spiders–Juvenile literature.
Classification: LCC QL458.42.A3 A43 2025 | DDC 595.4/4–dc23/eng/20240129
LC record available at https://lccn.loc.gov/2023059123

Portions of this work were originally authored by Amy Hayes and published as *Funnel-Web Spiders*. All new material in this edition is authored by Roxanne Aldolpho.

Designer: Andrea Davison-Bartolotta
Editor: Brianna Propis

Photo credits: Cover, pp. 1, 5, 11 (both), 15, 19 Ken Griffiths/Shutterstock.com; p. 7 NNphotos/Shutterstock.com; p. 9 (cartoon spider) Apolinarias/Shutterstock.com; p. 9 (map) Suriya S9/Shutterstock.com; p. 13 IJPhoto/Shutterstock.com; p. 17 Bartsch Photography/Shutterstock.com; p. 21 Hutch Photographics/Shutterstock.com.

Printed in the United States of America

CPSIA compliance information: Batch #CSSA25. For Further Information contact Rosen Publishing at 1-800-237-9932.

CONTENTS

A WEB OF SPIDERS 4
FLAT, FUNNELED WEBS 6
AROUND THE WORLD 8
MATING 10
STICKY FUNNELS 12
TARANTULAS 14
THE FEARED FAMILY 16
WHAT HAPPENS
IF YOU'RE BITTEN? 18
SCARY OR COOL? 20
GLOSSARY 22
FOR MORE INFORMATION 23
INDEX 24

Words in the glossary appear in **bold** type the first time they are used in the text.

A WEB OF SPIDERS

Spiders are fascinating animals! With their eight legs and—usually—eight eyes, they can look pretty creepy! There are many different types of spiders, and you might've seen some in a dusty corner of your house or lurking outside.

Spiders are put into different groups. These groups can be based on the size of spiders or the type of webs they **weave**. One kind of spider in particular weaves a special kind of web. Let's learn about the group known as funnel-web spiders.

There are many different types of spiders, but funnel-web spiders are very unique!

FUN FACT

MORE THAN 50,000 SPECIES, OR TYPES, OF SPIDERS HAVE BEEN DISCOVERED. SCIENTISTS MIGHT STILL UNCOVER NEW SPECIES OF SPIDERS LIVING ON EARTH TODAY.

FLAT, FUNNELED WEBS

A funnel is an upside-down, hollow cone. It's wide at the top and narrow at the bottom. Many people use a funnel in their kitchen to pour liquids.

Funnel-web spiders usually build a flat web to catch **prey** and a funnel-like tube leading to a hole where the spider hides. When a yummy meal—like a fly—lands on the flat web, the spider quickly grabs it and eats it in the funnel!

FUN FACT

FUNNEL-WEB SPIDERS MAKE VENOM INSIDE THEIR BODIES THAT PARALYZES PREY! THIS MAKES IT EASY FOR THE SPIDERS TO DRAG THEIR MEALS BACK TO THE CENTER OF THE FUNNEL.

The flat part of the web is like a spider's kitchen, and the funnel is their dining room.

AROUND THE WORLD

There are three families of funnel-web spiders. Agelenidae live throughout the world, including in North America. Dipluridae live mostly in Central and South America. Hexathelidae live in Australia.

Spiders in the Agelenidae family are nocturnal, which means they're active at night. You might see one in a **horizontal** funnel web in low bushes or grass. They also might make webs under rocks or in the corners of homes and sheds.

WHERE FUNNEL-WEB SPIDERS LIVE

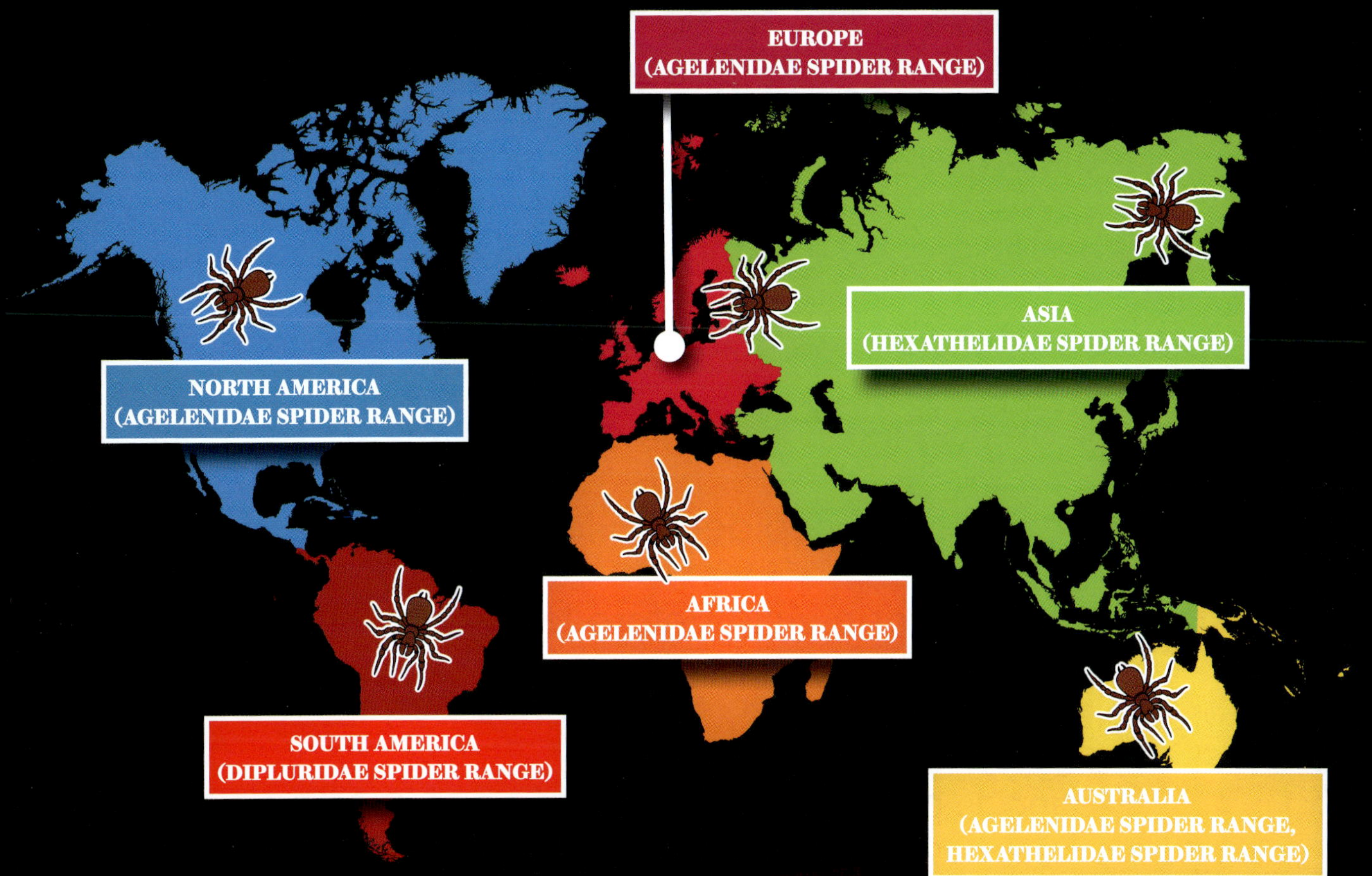

FUN FACT

SCIENTISTS BELIEVE NEARLY ALL AGELENIDAE SPIDERS ARE HARMLESS TO HUMANS. BUT IT IS POSSIBLE THAT THE VENOM OF THE *ERATIGENA AGRESTIS*, ALSO KNOWN AS THE HOBO SPIDER, MAY BE DANGEROUS.

MATING

Male funnel-web spiders don't spin webs. Instead, they wander around looking for females—who do spin webs—to **mate** with. After mating, the female will produce an egg sack. In some species, this sack is built inside the web. Baby spiders called spiderlings will stay in the web for awhile before leaving to begin life on their own.

Mating is dangerous for male funnel-web spiders. The female may try to kill and eat the male spider after or during the mating process!

FUN FACT

SOME FEMALE FUNNEL-WEBS PRETEND TO BE DEAD BEFORE OR DURING THE MATING PROCESS TO TRICK THE MALE!

Female funnel-web spiders are larger than males. But, males make a deadlier venom.

STICKY FUNNELS

The average life span of spiders in the Agelenidae family is one to two years. Many die during the winter. They can live longer in warmer areas. These spiders eat a lot of bugs and other creatures—including other funnel-web spiders!

Some funnel webs are sticky. If a funnel-web spider crawls across another spider's sticky web, it can get stuck. Then it becomes spider prey! But if you see a funnel-web spider, don't be scared. It will stay away from you if you stay away from it!

In Europe and America during the 17th and 18th centuries, funnel-web spider webs were used as bandages to stop bleeding.

FUN FACT

MOST OF THE AGELENIDAE SPIDERS ARE VERY FAST RUNNERS. IF YOU SURPRISE AN AGELENIDAE SPIDER, YOU MIGHT ONLY CATCH A GLANCE BEFORE IT QUICKLY HIDES!

TARANTULAS

Spiders in the Dipluridae family have another name: funnel-web tarantulas! They're also sometimes called curtain-web spiders. Their cloudy, messy webs fill spaces under rocks, logs, and even moss. They build many funnels that reach into these spaces rather than into holes in the ground.

Unlike the other funnel-web families, Dipluridae spiders live all around the world in warm, wet places such as rainforests. They also have longer spinnerets than many other funnel-web spiders.

Spinnerets are the **organs** that make a spider's silk, which is used for webs.

FUN FACT

SPIDERS IN THE DIPLURIDAE FAMILY RANGE IN SIZE. THE SMALLEST ARE ONLY 0.12 INCH (3 MM) LONG, BUT THE LARGEST MAY GROW TO 1.2 INCHES (30 MM)!

THE FEARED FAMILY

Of the three funnel-web spider families, the Hexathelidae family is the most well-known—and the most feared. This spider family lives in Australia and is known for its powerful venom. Black or brown and somewhere between 0.4 and 2 inches (1 and 5 cm), these spiders need lots of moisture. Small but deadly, they may be able to live up to 20 years!

These spiders mostly stay in their web, luckily for us. However, the males go out at night during the summer in search of a spider mate—so keep an eye out!

Many species of the Hexathelidae family live in burrows, or holes, with lines of silk running from them. Some species, such as *Hadronyche formidabilis*, live in trees, as seen here.

FUN FACT

A BITE FROM A MALE HEXATHELIDAE IS FIVE TIMES AS VENOMOUS AS A FEMALE'S AND CAN KILL A PERSON!

WHAT HAPPENS IF YOU'RE BITTEN?

The most dangerous spider in the world is considered to be the Sydney funnel-web spider, or *Atrax robustus*. This black or brown spider comes out after the rain, sometimes falling into swimming pools. It has been known to wander into homes to find a damp place to stay during the day.

If you get bitten by one of these spiders, you should get to a hospital as soon as possible! The bite will cause pain, then **numbness**. Scientists believe you could die within a few hours after being bitten.

Some other symptoms of being bitten include an upset stomach and throwing up, sweating, and tongue **spasms**.

FUN FACT

THIRTEEN PEOPLE HAVE DIED FROM SYDNEY FUNNEL-WEB SPIDER BITES. NO DEATHS HAVE BEEN RECORDED SINCE THE CREATION OF THE **ANTIVENOM** IN THE EARLY 1980S.

SCARY OR COOL?

Funnel-web spiders are as scary as they are amazing! Their unique webs can be found in burrows under rocks or all the way up in the treetops. They are found all over the world, with many different species to learn about.

Some funnel-web spiders are mostly harmless, while others are said to be among the most feared species of spiders. No matter which species you encounter, it's best to leave these eight-legged terrors alone.

Funnel-web spiders are scary to the bugs and animals that end up as their dinner. For people, they're best left alone.

FUN FACT

ONE OF THE WAYS TO KEEP FUNNEL-WEB SPIDERS FROM ENTERING YOUR HOME IS BY KEEPING MOISTURE OUT. ANY LEAKY PIPES, WET SINKS, AND DRAINING AIR CONDITIONER UNITS COULD DRAW THESE SPIDERS!

GLOSSARY

antivenom: A medicine, or drug, that stops a poison from working.

horizontal: Being longer side to side, with the top and bottom on the left or right.

mate: One of two animals that come together to produce babies. Also, to come together to make babies.

numbness: A lack of feeling in the body or a body part.

organ: A part of the body with a special function, such as the heart, kidneys, and lungs.

paralyze: To make unable to move.

prey: An animal that is hunted by other animals for food.

spasm: An involuntary muscle movement.

venom: Something an animal makes in its body that can harm other animals.

weave: To pass matter back and forth over itself to make a pattern.

FOR MORE INFORMATION

Books

Higgins, Melissa. *Splendid Spiders*. North Mankato, MN: Pebble, 2020.

Patterson, Jack K. *Spiders*. New York, NY: Cavendish Square, 2019.

Websites

National Geographic
www.nationalgeographic.com/animals/article/milking-venomous-funnel-web-spiders-australia-spd
Read about how a funnel-web spider's venom is used to make antivenom.

World's Deadliest Spider: the Funnel-Web
www.australiangeographic.com.au/topics/wildlife/2017/02/worlds-deadliest-spider-the-sydney-funnel-web/
Learn more about funnel-web spiders—and how to keep from being bitten!

Publisher's note to educators and parents: Our editors have carefully reviewed these websites to ensure that they are suitable for students. Many websites change frequently, however, and we cannot guarantee that a site's future contents will continue to meet our high standards of quality and educational value. Be advised that students should be closely supervised whenever they access the internet.

INDEX

Agelenidae, 8, 9, 12, 13

Australia, 8, 16

bite, 18, 19

Central America, 8

Dipluridae, 8, 14, 15

eggs, 10

eyes, 4

females, 10, 11, 16

food, 12, 21

Hexathelidae, 8, 16, 17

hobo spider, 9

life span, 12, 16

males, 10, 11, 16, 17

mate, 10, 16

North America, 8

prey, 6, 12

size, 11, 15, 16

South America, 8

spiderlings, 10

spinnerets, 14, 15

Sydney funnel-web spider, 18, 19

venom, 6, 9, 11, 16, 17

web, 4, 6, 7, 8, 10, 12, 13, 14, 15, 16, 20

where found, 8, 9